YOUR KNOWLEDGE HAS VALUE

- We will publish your bachelor's and master's thesis, essays and papers

- Your own eBook and book - sold worldwide in all relevant shops

- Earn money with each sale

Upload your text at www.GRIN.com and publish for free

Tackling Food Insecurity. A Case for Organic Agriculture

Akinmayowa Adedoyin Shobo

Bibliographic information published by the German National Library:

The German National Library lists this publication in the National Bibliography; detailed bibliographic data are available on the Internet at http://dnb.dnb.de.

ISBN: 9783346377753
This book is also available as an ebook.

TACKLING FOOD INSECURITY:
A Case for Organic Agriculture

Shobo Akinmayowa Adedoyin

2021

CONTENTS

Introduction 3

Organic Sources of Nutrients 6

Relationship between Food Security and Organic Source of Nutrients 8

References 16

Introduction

According to a Food and Agriculture organization (FAO) 2017 report on the future of food; the world today is increasingly confronted by multidimensional challenges from drought, food shortages, diseases, and global climate changes to an ever-increasing global population. As they are associated with one another, it is seen that to address some of these challenges, the issue of food security has to be tackled. To this effect, the agriculture industry thus faces an enormous task of increasing food production in an equitable and sustainable manner (Foley *et al.*, 2011; Stewart and Robert, 2012; Dinpnah, 2014; FAO, 2014; FAO, 2015; Ponisio *et al.*, 2015; Sarma, 2015).

There is a need to employ approaches that will enhance the resilience of agriculture production systems and contribute to mitigation of the current food insecurity through promoting biodiversity over mono-cropping, increasing soil organic matter and de-linking food production from fossil fuel reliance (De Schutter, 2010; Sarma, 2015; Bardgett and Gibson, 2017). In view of this, a balanced and precise crop nutrient application particularly of organic source has been suggested to be a prerequisite tool for meeting the second United Nations Sustainable Development Goal (SDG) to end hunger, achieve food security, improved nutrition and promote sustainable agriculture (International Fertilizer Association (IFA), World Farmers' Organisation (WFO) and Global alliance for climate-smart agriculture (GACSA), 2016).

According to a 2015 joint report by FAO, International Fund for Agriculture Development (IFAD) and WFP; one in nine people remain hungry in the world today. In fact, reports in various literature have shown that most developing countries notably sub-Saharan Africa constitute the "largest portion of world's chronically hungry people" (FAO, 2009; Holt-Gimenez *et al.*, 2012; Ponisio *et al.*, 2015; Sarma, 2015). Consequently, improved food production is being considered an integral solution in addressing hunger and resulting economic vulnerability.

In terms of agricultural capability and meeting demand for food; studies have pointed out that conventional agricultural practices such as the chemically-intensive, mono-cropping agriculture (amidst other current policies) may not even be sustainable and viable option for most farmers in the developing countries (Holt-Gimenez *et al.*, 2012); such argument of course relates to the small nature of the farmers' landholdings. This is among the myriad of reasons why there is growing interest for cost-economic, agro-ecological and organic forms of agriculture (De Schutter, 2010; Altieri *et al.*, 2012; IFAD, 2005; IFAD, 2003; FAO, 2017).

According to a report by International Federation of Organic Agriculture Movements (IFOAM), organic agriculture as a production system relies on "ecological processes, biodiversity, and cycles that are well-adapted to local conditions, rather than inputs with adverse effects." (IFOAM, 2006; Dinpnah,

2014; Sarma, 2015). It supports refraining from the use of synthetic or chemical based inputs to agricultural systems.

More importantly, organic agriculture views the farm somewhat as a living entity where all the components (soil minerals, organic matter, microorganisms, insects, plants, animals, and humans) interact with one another creating a self-regulating and stable system (IFOAM, 2006; Dinpnah, 2014; Sarma, 2015). Furthermore, it relies on natural regulating processes, local resources, and ecosystem capacities to optimize ecosystem functions, improve yields as well as disease resistance. This is in contrast to industrial agriculture, which relies heavily on non-renewable resources and prompts disruption of natural cycles drawing farmers into a vicious cycle of input use (to address pest/disease outbreaks and nutrient management (Sarma, 2015). Food is a basic need for survival; therefore, one cannot overstress the importance of maintain food security throughout the year. Some experts opine that a state of food security is attained when "all the general population at all times have both physical and economic access to sufficient, safe and nutritious food that meets their dietary needs for an active and healthy life." Food insecurity therefore exists when people lack access to sufficient amounts of safe and nutritious food, and therefore is not consuming enough for an active and healthy life. Researchers have attributed food insecurity to the '...unavailability of food, inadequate purchasing power, or inappropriate utilization at the household level' (Stewart and Robert, 2012; FAO, 2010a; Dinpnah, 2014; Morshedi *et al.*, 2017).

As discussed earlier, there are increasing statistics; the world population is expected to increase at exponential rates (Morshedi *et al.*, 2017). With this projection comes mounting concerns over an already higher than acceptable level of food insecurity (Stewart and Robert, 2012; Morshedi *et al.*, 2017). Consequently, in a bid to promote food security (with respect to optimal application of nutrients); a number of studies have recommended four principles namely; *right nutrient source* applied at the *right rate*, *right time*, and *right place.* These four dimensional approach have been proven to ensure appropriate use of nutrient resources and optimized productivity (FAO, 2010c; IFA, 2010; Stewart and Robert, 2012; FAO, 2014; Morshedi *et al.*, 2017).

Organic Sources of Nutrients

A large number of diverse materials can serve as sources of plant nutrients. Plants absorb nutrients from all sources however they take up nutrients only in their inorganic form. Organic nutrient sources must be mineralized (converted from an organic to an inorganic form) before being taken up by plants. The amount of nutrients provided by the different sources varies greatly between and within agro-ecosystems. Experts from organisations such as IFA, WFO, GACSA have stressed that sustainable crop nutrition can be achieved by identifying and utilizing all available sources of plant nutrients (IFA-WFO-GACSA, 2016). These can be natural, synthetic, recycled wastes or a range of biological products including microbial inoculants.

Accordingly, a certain natural supply of mineral and organic nutrient sources is present in soils; however, these often have to be supplemented with materials from exogenous sources for better plant yield. In practical farming, a vast variety of sources do indeed have agro-economic importance in spite of large differences in their nature, nutrient contents, forms, physicochemical properties and rate of nutrient release (Stewart and Robert, 2012; FAO, 2017).

Nutrient sources are generally classified as organic, mineral or biological. For the scope of this essay; organic source of plant nutrients was treated.

Organic nutrient sources are often described as *manures, bulky organic manures* or *organic fertilizers*. Most organic nutrient sources, have widely varying composition. For some, such as cereal straw; they release nutrients only slowly (owing to a wide carbon: nitrogen ratio). Nitrogen-rich leguminous green manures or oilcakes on the other hand decompose rapidly and release nutrients quickly (IFA-WFO-GACSA, 2016; FAO, 2017).

Briefly, according to the joint IFA, WFO, GACSA report; organic sources of nutrients are derived principally from substances of plant and animal origin (IFA-WFO-GACSA, 2016). Partially humified (derived from humus) and mineralized under the action of soil microflora, the organic sources act primarily on the physical and biophysical components of soil fertility. There include manures made from cattle dung, excreta of other animals, other animal wastes, rural and urban wastes, composts, crop residues and even green manures. Bulky organic manure is a term that encapsulates materials such as cattle dung, farm yard manure, composts, etc. because of their large bulk in relation to the nutrients contained in them.

Essentially, as a pre-condition for growth, health and the production of nutritious food, plants require essential nutrients (macro and micronutrients) in sufficient (Reetz, 2016) namely: carbon (C), hydrogen (H), oxygen (O), nitrogen (N), phosphorus (P), potassium (K), sulphur (S), magnesium (Mg), calcium (Ca), iron (Fe), manganese (Mn), zinc (Zn), copper (Cu), boron (B), molybdenum (Mo), chlorine (Cl), nickel (Ni), sodium (Na) and cobalt (Co). Therefore, if a single essential plant nutrient is available in insufficient quantity, it affects plant growth and invariably the yield.

Experts agree with Liebig's law of minimum that the level of plant growth is limited by the nutrient that is present in the environment in the least concentration relative to its demands for growth (Reetz, 2016). Certain essential nutrients such as (not limited to) N, P and K are generally assumed to be the most widely deficient elements.

Meanwhile, it is important to note at this juncture that in addition to organic sources, nutrients can be sources from rock weathering, atmospheric deposition, added irrigation water, crop residues, compost, livestock manure, bio-solids, and manufactured fertilizers (FAO, 2015; Reetz, 2016).

Relationship between Food Security and Organic Source of Nutrients

As emphasized earlier, food insecurity and an ever-expanding world population are a few instances of the challenges of the 21st century world. On one hand, providing food for the growing population requires a tremendous increase in the level of agricultural production (Dinpnah, 2014). On the other hand, given the importance of food security and the irreparable damage due to excessive use of agricultural chemicals, increased attention is currently focusing on safer, viable, cost-effective organic alternative (Morshedi *et al.*, 2017).

Organic farming is based on minimizing the use of external inputs, fertilizers, insecticides, and pesticides (FAO, 2015; WHO, 2015).

Accordingly, the evaluation of the benefits and limitations of organic agriculture appear to be complex given that its impacts depend on various factors such as the state of soil condition, farmer's knowledge and skills as well as resources availability (IFAD, 2015; Sarma, 2015).

Various scholarly works have cited the role of organic agriculture to developing countries to its economic, environmental, and social benefits (Mondelaers *et al.*, 2009; Peramaiyan *et al.*, 2011; Ward, 2013; Seufert, 2012; Bahramian and Mirdamadi, 2011; Omidi, 2014; Sarma, 2015

With respect to the **environmental benefits**; Organic agriculture provides greater environmental benefits per unit area than input-intensive industrial/ conventional systems on a variety of indicators including increased soil organic matter as well as soil organic carbon concentration and stocks, (Gattinger *et al.*, 2012) improved agronomic, natural, and soil biodiversity; reduced nitrogen and phosphorous leaching and greenhouse gases (GHG) emissions (Seufert, 2012; Mondelaers *et al.*, 2009; Tuomisto *et al.*, 2012; Morshedi *et* *al.*, 2017). There is also evidence that organic systems can contribute to soil carbon sequestration (Rodale institute, 2014) and higher sequestration rates compared to conventional systems. A meta-analysis in the context of developing countries has highlighted increased yield in organic production systems (Badgely *et al.*, 2007; Gibbon and Bolwig, 2007; Morshedi *et al.*, 2017).

Furthermore, Organic farming leads to healthier soils with reduced soil erosion (Morshedi *et al.*, 2017) through enhancement of soil organic matter. It has been shown to be instrumental to maintaining soil fertility (nutrients and soil diversity) and soil functions (better water retention, nutrient uptake, buffering and filtering capacities), providing ecosystem services like crop protection (IFOAM, 2006; Mondelaers *et al.*, 2009; Platteau, 2005; Pimentel *et al.*, 2005). Soil organic matter and quality have been associated with organic production techniques demonstrating higher or similar yields to conventional systems (Herencia *et al.*, 2008; Melero *et al.*, 2006).

Organic systems have shown to significantly improve species richness (Bengtsson, 2005; Hole, 2005; Morshedi *et al.*, 2017). This feature not only enables greater conservation of genetic diversity and protection from disease and pests, but also over the time improves resilience in the system (IFOAM, 2006).

Studies on organic agriculture also show increased soil water content, water retention, and volume of water percolation indicating improved ground water recharge and reduced runoff when compared to conventional systems (Pimentel *et al.*, 2005; Morshedi *et al.*, 2017).

In terms of energy requirement, organic farming requires lower energy input compared to the conventional systems and also high energy efficiency (Tuomisto *et al.*, 2012). Studies have also indicated a 6–30 % reduction in the global warming potential per kilo of product for organic products peaking at 41% (ITC-FiBL, 2007; Morshedi *et al.*, 2017).

During extreme weather conditions like drought, Organic systems have been demonstrated to perform better than conventional systems (Lotter, 2003; Gomiero, 2008; FAO, 2017). This is especially important in developing countries.

Economically, a number of factors such as yields, prices, and production costs (including cost for inputs, labour and certification) have been termed as important indices to measure productivity in organic agriculture costs (IFOAM, 2006; Seufert, 2012; FAO, 2017). First; yields can be influenced by variables such as production system characteristics, organic nutrient management, farmer's level of knowledge among others (Forster *et al.*, 2013; Sarma, 2015; Morshedi *et al.*, 2017).

Compared to traditional farming systems; organic farming practices demonstrate improved crop productivity and yield up to 170% (Badgley *et al.*, 2007). This has been documented in such systems in Africa, China, India, and Latin America (IFAD, 2005; Gibbon and Bolwig, 2007; Ponisio *et al.*, 2015; Morshedi *et al.*, 2017).

Researchers found for instance; that in addition to the use of leguminous crops, microbial nutrient fixers among others; manure and composts also possess a huge potential in improving yield performance (Rupela *et al.*, 2006; Chappell, 2007; Venkateswarlu, 2008; Seufert, 2012).

Long-term farm trial comparisons between organic and conventional systems in places like the United States (US) indicate that yields in organic and conventional systems can be matched in cases of crops like corn, soya bean, wheat (Pimentel, 2005; Rodale Institute, 2011). More significantly, organic farms achieve higher yields in comparison to conventional farms in drought conditions (Rodale Institute, 2014) out-yielding conventional crops by even 70–90% in severe conditions (Gomiero *et al.*, 2008; Sarma, 2015).

According to the IFOAM report (2006); 'productivity assessments' (or production system characteristics) should be based on total farm productivity rather than single crop yield estimates. Organic production systems have been shown to maximize productivity of the farm due to its capability for growing multiple crops compared to conventional methods driven by single crop farming (Seufert *et al.*, 2012; Serma, 2015; Ponisio *et al.*, 2015; Morshedi *et al.*, 2017).

Meanwhile, as highlighted above, the ability of organic nutrient management practices to adequately replenish soil nutrients, removed by crop harvests remains a contentious issue in farming systems (International Food Policy Research Institute (IFPRI), 2002). However, considering other variables of organic nutrient management; when compared to conventional systems;

organic agriculture have been shown to result in better pest and disease control mechanisms (Ponisio *et al.*, 2015). Additionally, soils in organic farms managed (with farm yard manure for instance) for long periods have been demonstrated to show higher content of organic matter and nutrients such as nitrogen that improves soil fertility (IFOAM, 2006; Pimentel, 2005; Sarma, 2015).

In terms of production costs; researchers found the organic agricultural systems to have significantly reduced the overall production costs in comparison to conventional production system (Eyhorn *et al.*, 2007; Valkila 2009; Panneerselvam *et al.*, 2011; Seufert, 2012; Sarma, 2015).

On its potential for *social impact*; organic agriculture is able to promote food security in various ways. For instance; for the farmers operating in traditional, small-scale or low-input systems, it improves farm yields and incomes. It leads to enhancement in food availability because of the possibility of diversification and mixed farming. There is also the case for lowering chances of crop failure in case of extreme climate events.

Resource-poor farmers especially in developing countries can also avoid taking high interest loans as organic production systems rely on locally available inputs rather than synthetic agricultural inputs which amongst other things increases production costs (IFAD, 2005; Sarma, 2015; Morshedi *et al.*, 2017). More so, organic agriculture can lead to diminished pesticide exposure amongst farmers and agricultural workers. Accordingly, some scholars have argued that when organic sources of nutrients (in combination with other organic agricultural practices) are employed; the challenge of crops containing pesticide residues and the emergence of antibiotic resistance bacteria will have been addressed (Smith-Spangler *et al.*, 2012; Morshedi *et al.*, 2017). This is an existing challenge in conventional farming that used synthetic chemicals in eliminating pests. Meanwhile, food from organic farming system have been reported to arguably contain higher levels of antioxidants and lower concentrations of toxic metal such as cadmium (Baranski *et al.*, 2014; Dangour *et al.*, 2009).

Furthermore, it is important also to highlight that this social impact of using organic sources of nutrients (within the framework of the organic agricultural system) extends to its ability to reinforce community relationships among farmers and other critical stakeholders such as non-governmental organisations within agricultural sector, and extension workers. There is also a case for better farmer-consumer relations in the local communities. More so, it provides innovative solutions on affordability as farmers from developing countries who generally find it difficult to access credit to purchase conventional inputs can now seek alternative options in organic modes of production (Tovignan and Nuppenau, 2004; Goldberger, 2008; Mendez *et al.*, 2010 Thapa and Morshedi *et al.*, 2017).

In conclusion, from various sources (as reviewed in this discourse); organic farming revered for benefits since time immemorial have been recommended to lead the direction of agriculture for the future. Its affordability, accessibility amongst other benefits makes it particularly an interesting option for both developing and developing economies in the fight against food insecurity in today's world. As global population peaks; as we discuss the issue of climate change in conjunction wars and insurgency which are amongst the leading world challenges; there is the need to give emergency attention to trends in food insecurity. This is therefore, a call for stakeholders to conduct more research and innovativeness for the purpose of achieving the United Nations SDGs of zero hunger and mitigating global food insecurity.

References

Altieri M, A., Funes-Monzote, F. R., and Petersen, P. 2012. Agroecologically efficient agricultural systems for smallholder farmers: contributions to food sovereignty. Agronomy for Sustainable Development 32: 1–13.

Badgley, C., Moghtader, J., Quintero, E., Zakem, E., Jahi Chappell, M., Avile´s-Va´zquez, K., Samulon, A., and Perfecto I. 2006. Organic agriculture and the global food supply, Renewable Agriculture and Food Systems 22(2): 86–108.

Bahramian, S.; Mirdamadi, S.M. Organic agriculture, increase farmers' income. J. Barzgar 2011, 2, 3–4.

Baranski, M. and Srednicka-Tober, D. 2014. Higher antioxidant and lower cadmium concentrations and lower incidence of pesticide residues in organically grown crops: A systematic literature review and meta-analyses. British Journal of Nutrition 112(5): 794–811.

Bardgett RD., Gibson DJ (2017). Plant Ecological Solutions to Global Food Security. Editorial: Ecological Solutions to Global Food Security. Journal of Ecology, 105, 859–864. Doi: 10.1111/1365-2745.12812.

Bengtsson, J., Ahnstrom, J., and Wei bull, A.C. 2005. The effects of organic agriculture on biodiversity and abundance: A meta-analysis. Journal of Applied Ecology 42(2): 261–269.

Chappell, M. J. 2007. Shattering Myths: Can Sustainable agriculture feed the world. Food first backgrounder. Institute for Food and Development Policy, 13(3).

Dangour, A. D. 2009. Nutritional quality of organic foods: A systematic review. The American Journal of Clinical Nutrition 90(3): 680–685.

De Schutter, O. 2010. Agroecology and the right to food. United Nations Office of the Special Rapporteur on the Right to Food.

Dinpnah, G.; Jamshid Nouri, A. Factors affecting feasibility of hydroponics cultivation based on infrastructure. J. Agric. Ext. Educ. Res. 2014, 26, 83–92.

Eyhorn, F. *et al.* 2007. The viability of cotton-based organic farming systems in India. International Journal of Agricultural Sustainability 5(1): 25–38.

FAO (2017). The future of food and agriculture – Trends and challenges. Food and Agriculture Organization of the United Nations, Rome. ISBN 978-92-5-109551-5.

FAO 2010. FAOSTAT. Food and Agriculture Organization of The United Nations Statistics. [Online]. Available at Http://Faostat.Fao.Org/ (accessed 8 April 2011).

FAO, IFAD and WFP. 2015. The State of Food Insecurity in the World. Meeting the 2015 international hunger targets: Taking stock of uneven progress. Rome, Food and Agriculture Organization of the United Nations.

FAO. 2009. How to Feed the World: 2050 [Online]. High Expert Forum, 12-13 Oct. 2009. Food and Agriculture Organization of the United Nations. Rome. Available At Http://Www.Fao.Org/Fileadmin/Templates/Wsfs/Docs/Issues_Papers/HLEF2 050_Global_Agriculture.Pdf (accessed 31 March 2020).

FAO. 2010a. Basic Definitions [Online]. Food and Agriculture Organization of the United Nations. Rome. Available at Http://Www.Fao.Org/Hunger/Basic-Definitions/En/ (accessed 30 March 2020).

Foley, J. A. *et al*. 2011. Solutions for a cultivated planet. Nature 478: 337–342.

Food and Agriculture Organization (FAO) (2014). Towards Sustainable Agriculture and Improved Food Security & Nutrition. Bangladesh Country Programming Framework CPF 2014 – 2018.

Food and Agriculture Organization of the United Nations (FAO) (2015). An Introduction to the Basic Concepts of Food Security. Available online: http://www.fao.org/docrep/013/al936e/al936e00.pdf (accessed on 16 February 2020).

Forster, D., Andres, C., Verma, R., Zundel, C., and Messmer, M. M. 2013. Yield and Economic Performance of Organic and Conventional Cotton- Based Farming Systems – Results from a Field Trial in India. PLoS ONE 8(12): e81039.

Gattinger, A., Muller, A., Haeni, M., Skinner, C., Fließbach, A., Buchmann, N., Mäder, P., Stolze, M., Smith, P., El- Hage Scialabba, N., and Niggli, U. 2012. Enhanced top soil carbon stocks under organic farming. PANS Early Edition.

Gibbon, P., Bowling, S. 2007. The economics of certified organic farming in tropical Africa: A preliminary analysis. SIDA DIIS. Working paper, no 2007/3, Subseries on Standards and Agro food exports (SAFE) no 7.

Goldberger, J. R. 2008. Diffusion and adoption of noncertified organic agriculture: a case study from semi-arid Makueni District, Kenya. Journal of Sustainable Agriculture 32(4): 531–564.

Gomiero, T. 2008. Energy and Environmental Issues in Organic and Conventional Agriculture. Critical Reviews in Plant Sciences 27(4): 239–254.

Herencia, J. F., Ruiz, J. C., Melero, S., Galavis, P. A. G., and Maqueda, C. 2008. A short-term comparison of organic v. conventional agriculture in a silty loam soil using two organic amendments. Journal of Agricultural Science 146: 677–687.

Hole, D. G. *et al*. 2005. Does organic farming benefit biodiversity? Biological Conservation 122(1): 113–130.

Holt-Giménez, E., Shattuck, A., Altieri, Miguel., Herren, H., and Gliessman, Steve. 2012. We already grow enough food for 10 billion people and still can't end hunger. Journal of Sustainable Agriculture 36(6): 595–598.

IFA, WFO and GACSA (2016). Nutrient Management Handbook.

IFA. 2010. International Fertilizer Industry Association Statistics. [Online]. Available At Http://Www.Fertilizer.Org/Ifa/Home-Page/STATISTICS (Verified 8 April 2011).

IFAD. 2003. The adoption of organic agriculture among small farmers in Latin Americans and Carribean thematic evaluation, International Fund for Agriculture Development (IFAD) Report No 1337.

IFAD. 2005. Organic agriculture and poverty reduction in Asia: China and India focus, thematic evaluation, International Fund for Agriculture Development (IFAD) Report No 1664.

IFOAM. 2006. Organic Agriculture and Food Security, Dossier, International Federation of Organic Agriculture Movements.

International Fund for Agricultural Development (IFAD) (2015); World Federation of Exchanges (WFE); Food and Agriculture Organization of the United Nations (FAO). The State of Food Insecurity in the World Food Insecurity in the World. Available online: http://www.fao.org/docrep/018/i3434e/i3434e00.htm (accessed on 12 May 2020).

ITC – FiBL. 2007. Organic Farming and Climate Change, International Trade Centre UNCTAD/WTO, Geneva: ITC, 27pp.

Lotter, D. 2003. The performance of organic and conventional cropping systems in an extreme climate year. American Journal of Alternative Agriculture 18(3): 146–154.

Melero, S., Porras, J. C. R., Herencia, J. F., and Madejon, E., 2006. Chemical and biochemical properties in a silty loam soil under conventional and organic management. Soil and Tillage Research 90: 162–170.

Méndez, V. E. *et al*. 2010. Effects of fair trade and organic certifications on small-scale coffee farmer households in Central America and Mexico. Renewable Agriculture and Food Systems 25(3): 236–251.

Mondelaers, K., Aertsens, J., and van Huylenbroeck, G. 2009. A meta-analysis of the differences in environmental impacts between organic and conventional farming. British Food Journal 111(10): 1098–1119.

Morshedi L., Lashgarara F., Hosseini SJF. Najafabadi M.O. (2017). The Role of Organic Farming for Improving Food Security from the Perspective of Fars Farmers. MDPI, Sustainability. 9, 2086; doi:10.3390/su9112086. www.mdpi.com/journal/sustainability.

Omidi N M.A. gender sensitive analysis towards organic agriculture: A structural equation modeling approach. J. Agric. Environ. Ethics 2014, 27, 225–240.

Panneerselvam, P., Hermansen, J. E. 2011. Food Security of Small Holding Farmers: Comparing Organic and Conventional Systems in India. Journal of Sustainable Agriculture 35(1): 48–68.

Peramaiyan, P.; Halberg, N.; Hermansen, J.E. Food security of small holding farmers comparing organic and conventional in India. J. Sustain. Agric. 2011, 1, 48–68.

Pimentel, D., Hepperly, P., Hanson, J., Douds, D., and Seidel, R. 2005. Environmental, Energetic, and Economic Comparisons of Organic and Conventional Farming Systems. BioScience 55(7): 573–582.

Pimentel, D., Hepperly, P., Hanson, J., Douds, D., and Seidel, R. 2005. Environmental, Energetic, and Economic Comparisons of Organic and Conventional Farming Systems. BioScience 55(7): 573–582.

Platteau, J., Bas, L., Bernaerts, E., Campens, V., Carels, K., Demuynck, E., Hens, M., Overloop, S., Samborski, V., Smets, D., Van Gijseghem, D., Vriesacker, M., and Wustenberghs, H. 2006. Landbouwbeleidsrapport 2005 (LARA). Brussel, Administrative, Department Landbouw en Visserij, Afdeling Monitoring en Studie, D/2006/3241/155, pp 240.

Ponisio, C. *et al*. 2015. Diversification Practices reduce organic to conventional yield gap. Proceedings of The Royal Society B 282: 21041396.

Reetz, H.F. Jr. (2016). Fertilizers and Their Efficient Use. International Fertilizer Industry Association, Paris, France.

Rodale Institute. 2014. Regenerative Organic Agriculture and Climate Change: A Down-to-Earth Solution to Global Warming, United States.

Rupela, O. P., Gowda, C. L. L., Wani, S. P., and Bee, H. 2006. Evaluation of Crop Production Systems Based on Locally Available Biological Inputs, In Biological Approaches to Sustainable Soil Systems, pp 501–515.

Sarma SD (2015). Organic Agriculture: An option for fostering sustainable and inclusive agriculture development in India: Discussion Paper. The Energy and Resources Institute. pp 1- 28. www.terrin.org.

Seufert, V. *et al.* 2012. Comparing the yields of organic and conventional agriculture. Nature 485: 229–232.

Smith-Spangler, C., Brandeau, M. L., Hunter, and Grace E. 2012. Are Organic Foods Safer or Healthier Than Conventional Alternatives? A Systematic Review. Annals of Internal Medicine. 157(5):348–366.

Stewart W.M., Roberts T.L. (2012). Food Security and The Role of Fertilizer In Supporting It. 1st International Symposium On Innovation And Technology In The Phosphate Industry. Procedia Engineering 46: 76 – 82.

Tovignan, D. S. and Nuppenau, E. A. 2004. Adoption of organic cotton in Benin: Does gender play a role. Conference on Rural Poverty Reduction through Research for Development and Transformation, Berlin.

Tuomisto, H. L., Hodge, I. D., Riordan, P., and Macdonald, D.W. 2012. Does organic farming reduce environmental impacts? –A Meta-Analysis of European research. Journal of Environmental Management 112: 309–320.

Valkila, J. 2009. Fair trade organic coffee production in Nicaragua-- Sustainable development or a poverty trap? Ecological Economics 68(12): 3018–3025.

Venkateswarlu, B. 2008. Organic Farming in Rain-fed Agriculture: Prospects and Limitations, Chapter 2, pp 7–11.

Ward, C.; Reynolds, L. Organic agriculture contributes to sustainable food security. Vital Signs 2013, 20, 66–68.

World Health Organization (WHO) (2015); Food and Agriculture Organization of the United Nations (FAO). Organically Produced Foods. Available online: http://ftp.fao.org/docrep/fao/010/a1385e/a1385e00.pdf (accessed on 14 June 2020).

YOUR KNOWLEDGE HAS VALUE

- We will publish your bachelor's and
  master's thesis, essays and papers

- Your own eBook and book -
  sold worldwide in all relevant shops

- Earn money with each sale

Upload your text at www.GRIN.com
and publish for free